INSTAGRAM MARKETING

GAIN MILLIONS OF FOLLOWERS AND MONETIZE YOUR INTAGRAM ACCOUNT

Elizabeth Nightingale

Table of Contents

Introduction 6
What is Instagram and Why Should I Use It? 8
The Benefits of Marketing with Instagram 12
The Difference Between Instagram and Other Social Media Sites 19
How to Create Your Own Business Account on Instagram 24
Starting Your Instagram Presence 27
Getting More Followers without Having to Spend Money 33
How to Market the Brand and Increase Sales 37
Getting Your Pictures to Work Well on Instagram 45
How to Successfully Interact with Others on Instagram 50
Common Mistakes to Avoid When Marketing on Instagram 55
The Best Tips for Growing Your Instagram Business Account 60
Conclusion 67

© Copyright 2017 by Elizabeth Nightingale *- All rights reserved.*

The following eBook is reproduced below with the goal of providing information that is as accurate and as reliable as possible. Regardless, purchasing this eBook can be seen as consent to the fact that both the publisher and the author of this book are in no way experts on the topics discussed within, and that any recommendations or suggestions made herein are for entertainment purposes only. Professionals should be consulted as needed before undertaking any of the action endorsed herein.

This declaration is deemed fair and valid by both the American Bar Association and the Committee of Publishers Association and is legally binding throughout the United States.

Furthermore, the transmission, duplication or reproduction of any of the following work, including precise information, will be considered an illegal act, irrespective whether it is done electronically or in print. The legality extends to creating a secondary or tertiary copy of the work or a recorded copy and is only allowed with express written consent of the Publisher. All additional rights are reserved.

The information in the following pages is broadly considered to be a truthful and accurate account of facts, and as such any inattention, use or misuse of the information in question by the reader will render any resulting actions solely under their purview. There are no scenarios in which the publisher or the original author of this work can be in any fashion deemed liable for any hardship or damages that may befall them after undertaking information described herein.

Additionally, the information found on the following pages is intended for informational purposes only and should thus be considered, universal. As befitting its nature, the information presented is without assurance regarding its continued validity or interim quality. Trademarks that mentioned are done without written consent and can in no way be considered an endorsement from the trademark holder.

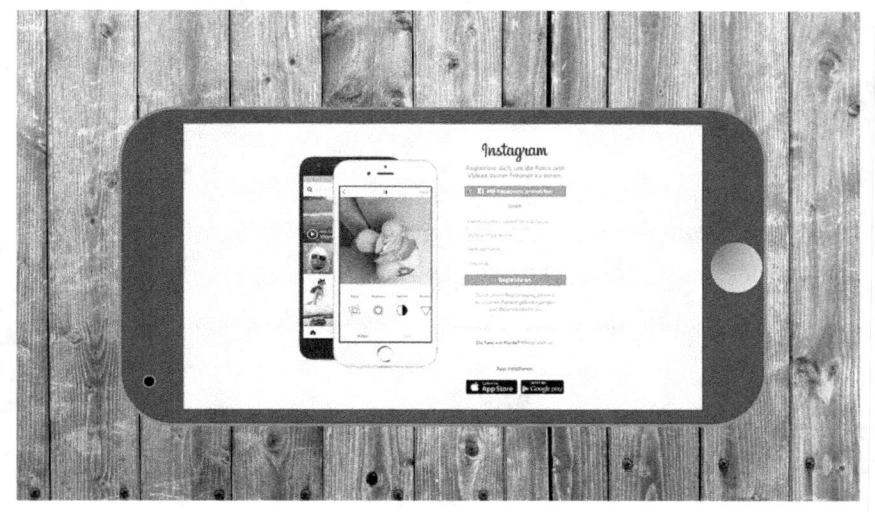

Introduction

Congratulations on downloading your personal copy of Instagram Marketing: Social Media Marketing with Instagram, Gain More Followers and Monetize Your Instagram Account. Thank you for doing so.

The following chapters will discuss some of the many great steps that you can take to start your own business account on Instagram, and how to use this account to grow your business. Marketing on Instagram is great for businesses because it relies on pictures and videos, something that many customers enjoy looking at when it is time to make a decision on one of their purchases. You can fill this need with your potential customers by setting up a professional looking account with some great media that showcases your work.

This guidebook is going to help you get the most out of your Instagram account. We will talk about some of the benefits of using Instagram over some of the other social media accounts for growing your business, as well as how you can set up your own account. We also work on how to increase your presence on social media, how to get more followers on Instagram without having to pay an expensive professional, and some great tips to help you pick out the perfect pictures to entice more customers.

Instagram is the perfect social media site to help businesses to grow. And with some of the great tips that are inside this guidebook, you will be able to get your Instagram account off the ground while growing your business in no time.

There are plenty of books on this subject on the market, thanks again for choosing this one! Every effort was made to ensure it is full of as much useful information as possible. Please enjoy!

What is Instagram and Why Should I Use It?

Before we get too far into marketing with Instagram, it is important to understand what Instagram is and why you would want to use it instead of another social media site in order to grow your business. Instagram is an app for social networking that was designed for sharing videos and photos from your smartphone. It does have some similarities to Twitter and Facebook in that when you create your own account on Instagram, you are going to have a profile as well as a news feed, but this one is really great for many businesses because it allows you to really showcase your business in a way that you just can't do with the other social media websites.

When you are on Instagram and you post a video or a photo, it is going to show up on your profile. The other people who are following your account (we will discuss later how to get some more followers) will be able to see these new posts on their own feed. You will also be able to choose some followers to watch and when they post, you can see their information as well.

This is a pretty simple social media site to work with and some see it as a simplified version of Facebook, but it will focus mostly on visual sharing and using your phone rather than long posts and blogs. You will be able to use your account to interact with some of the other users who are on

Instagram simply by following them, having them follow your account, private messaging, tagging, liking, and even commenting.

There are a lot of great devices running either Android or iOS that you can use in order to set up your free account with Instagram. In some cases, you can access the account from your own personal computer, but a lot of users like working with this social media site because they are able to use their smartphones in order to upload and share their media.

Before you are able to use this app, Instagram is going to ask you to create one of their free accounts. You can choose to sign up with your email account or with your Facebook account. You just need to come up with the username and password. While you are setting up the account, you may have a place where you will be asked if you want to follow some of the friends who are on your Facebook network on your Instagram one. You can choose if you want to do this now or skip it and do it later. (Or not at all!)

During this point, take some time to customize the profile by adding in a photo, a website link, a short bio, and your name. When you start looking for people who are going to follow you, or you decide to follow other people, they will want to know who you are so getting the profile set up will be a good idea.

As we mentioned above, you will want to use Instagram as a visual sharing site because this is the main purpose of using Instagram. You will want to find some of the best pictures and videos for your business and post these on your profile. You will also notice that every profile will also have a count that helps you to see who you are following and who is following you so that you can keep track of these numbers and make changes if you need. If you have someone who wants to follow you, they just need to tap on the Follow button and they can see what is on there. If you are a business, don't put the account to private or you will have to individually approve the request for all of them first.

An important part of working on Instagram, in addition to posting different media that works with your business, is to do interaction on posts and this can be really easy and a lot of fun. You can go on any post that you find and like and add a comment to it at the bottom, just make sure that your comments are well thought out and include more than just a few words. You can even use the arrow button in order to share it with someone else with the help of direct messaging.

The major thing that you will be doing with Instagram is sharing pictures, as well as some videos, that showcase your services or your products. What you post is going to vary based on what kind of business you are running, but you should make sure that the posts have value to your customers and that they really work with your business. Posting things from your recent vacation or from the meal that you had should be saved for your personal page, your business page is for your products and services and the posts should relate to this.

This can be a bit hard to think of when you get started, especially if you are just offering one service or just a few products that don't add in a lot of variety. Take a look through some other business pages that are similar to yours or in the same industry, and see what they used and what worked for them. This can at least get you started on some ideas and going in the right direction. You will want to change some of these around so that they are unique for your business, but it will really help you to get a good start.

If you are interested in finding some friends or even other interesting accounts to work with, you just need to use the search tab. It may not seem important to find other people on Instagram when you are a business, but finding these people to follow and commenting on their items and interacting with them can help other potential followers find you as long as you do all this work properly.

If you are looking to grow your business in a format that allows your customers to really get a good look at your products and services and that can be very interactive, then Instagram can be one of the best platforms that you will be able to utilize to really help you meet your own goals.

The Benefits of Marketing with Instagram

So why would you want to use Instagram for your marketing needs? Is it really that much better than some of the other social media sites, or are you just going to get on and find out that you are wasting your time or having to spend a lot more to get the results that you want? The good news is that there are many benefits to using Instagram for your business including the following:

Connect with your customers

When you think about connecting with your customers, you probably think about Facebook or maybe even Twitter to help you with that. But actually, Instagram is one of the most active when it comes to the user base and it is more likely that your followers will engage with you, interact, and even make purchases on Instagram compared to some of the other social media sites. In fact, a report that was done by Forrester Research found that your engagement can go up 120 times per follower compared to Twitter.

Businesses want to get communication and connections with your customers. They don't want to waste time doing all the work to just get a few people to respond to them. But with the high engagement that comes with Instagram, it is the best platform that you can use in order to connect with your customers, build relationships, and even listen to the feedback that you are given. While the other social media sites can work as well, they don't bring out the amount of engagement and potential sales as you can find with Instagram.

Learn what others like

If you have a business account set up already, your audience and followers could already be sharing the photos that you post and they could already be talking about your business on Instagram. This seems to be even more true for the brick and mortar businesses that have customers coming to their store on a regular basis.

For example, restaurants do well on places like Instagram because people like to share pictures of a new or favorite dish. Instagram makes it easy for them to do this because they can share that photo that they liked as well as allowing the customers to let others know where that meal is from. Restaurants are just one example of how this can work. You just need to post things that others will want to share and make sure that they are tagging your location as well. If you do this the right way, you will be able to get others to do some of the marketing for you.

If you would like to see if others have shared any of your photos and tagged your location is to do this as well. After you share the photo, you should be able to see a link that comes right above the photo and you will be able to click that link. This link is going to help you to see all of the photos that are shared from the office or the store and gives you a good idea of what some others are looking at in your business.

This is a good way to see which of your items are popular so you can make changes and pick out a new marketing plan. For example, if you have an item that is not doing the best, it may be a good idea to just skip on making that and focus on the items that do a little bit better so that you can get more shares and likes, and therefore more followers and sales, in the long run.

Reach a new audience

Instagram is set up to make it easier for the users to be able to discover new businesses, new people, and even new photos. This helps you out as a new business, but you need to make sure that you are taking the right steps to get other followers to find you.

One method that you can use to make people find you better is to use hashtags. This is similar to what you are able to do with some of the other social media sites. When you create a hashtag, you are adding a link to your pictures and videos. You can click on this link and view all of the other media that has been shared through Instagram with that hashtag. This allows you to become more readily available to the potential customers; when they place this hashtag into a search or some options that are similar, they will be able to see some of your media.

You as the business owner are able to use the hashtag as well in order to find some of your potential customers. You can search around for some of the keywords that you want and then meet some people who are sharing photos in your interest before making the introduction.

In addition to using these hashtags, Instagram has a cool think that is the Discover tab. This tab is dedicated to helping the various users to find photos that they want as well as helping them to connect with other users, or even other businesses, that are relevant to their needs.

Work with the other marketing channels

One thing that you will notice is that when you create some content on Instagram, even if you used some of their filters or other stuff to make the picture look better, you will be able to access and share it through all the different channels that you have for your marketing plan.

Within the settings inside of Instagram, you will be able to enable it to share your content on Twitter and Facebook. This means that if you upload some new photos to Instagram, they are automatically going to upload to your Twitter and Facebook accounts as well. you can even wait to share to the different networks until later, so that you use different release times, by saving the picture to the Camera Roll so that you can access them again later when you want to post.

In addition to using this content on your other social media accounts, you can also use it on your email marketing campaigns. Your settings inside of Instagram can be changed so that you can send some of your pictures and videos can be shown through email and other options to help make your marketing all work together.

Instagram and Facebook are created by the same company so you will find that these two in particular are able to work really well together. And since these two have the highest engagement and sales conversion out of all the social media platforms, it makes sense to combine them together to get even more sales for your business.

Generating sales

The main reason that you are on Instagram and other social media platforms is to generate more sales for your business. If you are able to create professional-looking images, you are going to have a better chance at promoting your product or even highlighting your services. and since you are able to use many of these social media sites for free (you can pay for a professional but often they are just going to do the same things that we talk about in this guidebook), this is a really affordable option to use.

Because all of this comes together, visual platforms, including Instagram are really successful at not only generating engagement, they are also able to drive up sales. And Instagram is able to bring in the sales more than some of the other options. One report from Shopify found that from Instagram, the average price for a sale on that site was $65. On the other hand, this average price is just $55 for Facebook and $46 for Twitter. This means that for the amount of work that you are doing, you could end up with much higher sales if you decide to use Instagram for at least part of the marketing campaign.

Of course, you need to make sure that you pick out the right media to post. You shouldn't fill up the feed with photos that have the same caption that just says "shop online" or "buy now" or something like that. Most people know that if you are a business posting a picture, you are trying to sell that item so find a way to be more creative and get their attention. Find cool and creative ways that you can display the products and then let the images do the speaking for you rather than using so many repetitive and boring sales tactics all the time. Telling a short story, using the right hashtags, and providing a quick link to your website in case the user is interested will be enough to get your sales.

Working with Instagram is not the only thing that is able to help you to see your sales go up, but when it comes to social media sites, this one is going to provide the highest response and conversion rates compared to other options like Twitter and Facebook. It is fine to use these as well, but if you want to start a marketing campaign with social media, it is definitely worth your time to add in Instagram to the mix.

There are many reasons that you should choose to go with Instagram to help you grow your business. It has higher engagement with the customers, which does translate into more sales, and will help you to get the results that you want. While it shouldn't be the only place that you do your marketing, it is a good place to get started to help your business soar.

The Difference Between Instagram and Other Social Media Sites

There are so many different social media sites that are out there and all of them promise to bring you the customers and sales that you want. But all of them are going to work in slightly different manners, so sometimes it depends on the kind of customer that you are trying to reach when it comes to which of the social media sites that you would like to work with. Let's take a look at how Instagram compares to some of the other social media sites that you may have used in the past and see how it can make a difference in your business.

Facebook

Facebook is an online social networking service that was launched in 2004. It allows you to meet up with friends and family and share pictures, videos, and blog posts about what is going on in your life. It is possible to have your own personal page on this site or you can create a new page that is professional and just for your business. This social media site can be accessed by tablets, smartphones, computers, laptops, and desktops. You can create a profile for Facebook and then indicate some information about yourself or your business so that others are able to find you.

Often businesses are going to start their own business sites, which you can do through a separate account or have it attached to your personal account so they are all in one place. You can share this information on *your marketing materials so that others can find you and keep up to date on what is going on in your business. This is a great place to be interactive with your customers, using videos, promotions, and pictures along with your blog posts to share information.*

Facebook can do quite a few of the same things that you find on Instagram, but it often relies more on text and blog posts than you will find with Instagram. You should still make sure that your posts are interactive and fun for your customers and take the time to fill out your business profile, complete with a good name and picture for your customers to find you.

Twitter

Twitter is another great social media site that businesses have come to use to help themselves grow. This one was launched in 2006 and as of 2012, there were more than 100 million users on this site, which provides you with many followers who may be interested in your information. this is a great website if you want to reach your customers in a quick and efficient manner, but you will be limited to writing out posts that are 140 characters or less.

This one is good if you want to post links to your website or other useful information, but you need to be creative order to fit into the small character count that is allowed. You can add in some pictures, but this is not the most effective method for posting pictures of your business. If you want a social media site that is really picture oriented and great for showcasing the products and services that are available on your website, this is not the one for you.

Pinterest

Pinterest is another photo sharing website that is going to rely quite a bit on pictures. The founders of this company state that Pinterest is like a catalog of ideas rather than having it there for social networking. This is a good site to use to really showcase the pictures that you have in pins on your different boards. You can take high-quality pictures of your products with some little descriptions. You can add in your website so that people can go there to check out more information and decide whether or not to purchase the products.

With this one, you do need to come up with some high-quality pictures to make others want to choose you over someone else. You need to have some good keywords in place so that your potential customers are able to find you and look at your products to purchase them.

The benefit of Instagram over Pinterest is that right now Pinterest doesn't really have the capabilities for videos that you can find with Snapchat and Instagram. There are a lot of great things that you can do with pictures on this one, but Instagram has more of the conversion rate that you want compared to Pinterest and since you are able to add in some videos on occasion to make it stand out, Instagram is often the best one that you can use.

Snapchat

Snapchat is a bit different because it is more about multimedia and image messaging. One of the ideas behind this social media app is that the pictures and the messages are going to be available for just a short period of time because they can't be found any longer. This one can be a little bit tough for some businesses because you have to send out messages that have a lot of power behind them that only last a few seconds before they are gone. This means that they don't have any staying power; you won't be able to keep showing them over and over again or have your customers share the information to become viral. But it can be a unique method of social media that you can use if you need something a little bit new.

Most business marketers are going to choose to skip out on this one even though there are many customers who are on it. It is kind of hard to create content all the time when it is going to just disappear. You will want most of your messages to stick and to have the ability to be shared with others across the media site. But if you create a message and then it is deleted shortly afterward, it is hard to have a lot of people see it. Most marketers want some more staying power and this means that it is best to go with some other options on social media.

Instagram

As you can see, some of these social media accounts will be similar to what you can do with Instagram. Some of them allow you to share pictures while others are better for doing videos or blogging. But with Instagram, you are focusing on creating content that is going works well in picture and video format. The better quality you are able to get these media on your account, the easier it is to attract your customers and get them to at least take a look at your site rather than just passing you by.

There are many great social media accounts that you are able to pick but when it comes to growing your business and reaching your customers in a way that really works to convert their views into sales, Instagram is one of the best options that you can choose.

has the option to use analytics so that you can take a look at who your followers are and how they behave.

Make sure to fill out all the different parts that come with your Instagram business account. There are so many aspects that can come into play with this and the more information that you post, the easier it is going to be to show your customers what you are all about. Fill out the industry and the bio, make sure that you put in the link to your website, and post some high-quality content so that others are easily able to find you.

Remember that with this account, you need to conduct yourself as a business. Make sure that you are posting things that relate to your business and the products or services that you sell. Make sure that you leave the personal stuff out of the mix and learn how to interact and form the right relationships. It is going to take some time to build all of this up and start to get some of the results that you want, but it is just like some of your other social media accounts and you just need to keep going to get the results that you want.

Starting Your Instagram Presence

Social media is one of the best ways to reach your target audience. It seems like everyone is on social media now and the companies that refuse to have any sort of presence on these sites are the ones who usually see a large decrease in their customer base. There are many social media sites that you can choose from and with a little bit of research, you will be able to find the one that has most of your target audience and use this to your advantage.

One of the social media sites that you may want to consider is Instagram. This is one of the most popular social networking sites and it is pretty simple to use and can showcase your business in many different ways. This social media site is going to allow you to share any video or photo that you would like with the potential to reach people all throughout the world. Many individuals choose to go on Instagram and use it to show what they are doing and to keep up with friends and family. But it can really provide some unique ways to showcase your business by showing off your service or product.

Think about the many ways that you would be able to use Instagram to boost your business. You can create videos that could go viral, put pictures up of new products, and share information in ways that are not possible with some of the other social network sites.

Posting on Instagram is the easy part, though. Like with other social networking sites, you will find that gaining some fans on Instagram can be tough, especially if you are new. Some businesses try to get dominance in this site by working on spam as well as other inappropriate content, but this is often going to make people annoyed with your page and you will start to lose some of the customers that you want. There are several options that you can choose from that will help you to find fans, fans who are there because they like your product or because they have purchased it in the past. Some of these options include:

Figure out what aspect you should focus on

When you want to do some marketing of your business with the help of Instagram, you need to figure out what you want to focus on. Many times beginners will want to try and dabble in everything and they hope that these activities will catch on so that others will start to follow them. This approach is not really a great idea because you are all over the place and really just wasting your time. So instead of trying to offer everything to your customers, choose one aspect and then concentrate on that one aspect.

Let's look at an example of how this would work. Say that you were creating an Instagram account in order to show off some of your painting skills. Here you will need to focus on just your skills in painting so make sure that all of the videos and photos that you work on should just show this skill. Yes, there are a lot of great memes and videos out there that are pretty popular, but they aren't really helping to promote the painting part of the business and if you add in all of those other things, you are just confusing your customers.

Remember that you are a business so you should be concentrating on your business. Unless you are a celebrity or you are using Instagram as your personal site, you should not update your customers on all your daily activities or post other things that have nothing to do with your business. Your customers are going to be there because they enjoy what you have to offer in services and skills, not because they want to know what you are having for supper.

Change the appearance of the account

After you pick out the focus that you want to work with, it is time to change around the appearance that goes with your account and make it look better. You should start out with a good profile picture as well as a good and thoughtful description that can help out with this task. This is a good start, but you need to do more if you want to stick out. It is important that you also secure the look of the photo feed of the account.

A simple way that you can do this is to collect about twenty pictures that go along with your chosen focus. This is going to be your portfolio and it should have some higher quality images. Make sure that if you have been using the Instagram page before that you get rid of all the stuff that is irrelevant or unattractive from the page. The photo feed in your page is a great way to attract others to the page so make sure that you are able to take high quality and interesting pictures or have someone else do it for you.

Share the Instagram page

Sometimes the best way to make sure that people are able to find your Instagram account is to share it with others. You can start with some of your other social networking sites, especially if these are better established for your business. Starting with Google+, Twitter, and Facebook, you can work through some of your social media sites and let others know that you have the Instagram account available.

In addition, if you have some marketing materials for your business or you have an email contact list of customers, you can send information about your Instagram to them. Some of them may be really interested in following you to find out more about the business, your promotions, and your products and services and this can be a great way to get ahold of some of your customers who may not be using the other sites you are on.

Interact on Instagram

By going through your existing network, you may be luck enough to get some great Instagram followers, but it is important to keep on working from here. If you are only dealing with your current fan base, you are not going to be getting anyone new and you won't see the growth that you want form your business. One thing that you are ale to do in order to get some of these followers is to interact with some of the other users on Instagram and search through their followers to see who is in their potential audience.

You need to make the interactions with the other Instagram users look as normal as possible; otherwise, it is going to show up as spam inside of the system. For example, don't use too many hashtags in the work, don't ask the user for a shout out, and don't do any exchanging of follows and likes. They were once effective when Instagram was brand new, but they are just not effective any longer.

The most effective method that you can use in order to get the attention of the user is to comment on a video or a photo that they uploaded. Make sure that the comment is positive and thoughtful, something that is more than a few words. You should leave your actual thoughts about the content and say something that will grab their attention. If you write out something that is thoughtful and that others will enjoy, you are more likely going to get others back to your page. Never write out that you want others to follow your account, though; this is seen as spam and will turn many potential fans away.

Engage the followers

Once you have some of the followers that you want on your page, it is not enough to just have them there, you need to be able to engage them as well. If you take these followers for granted and don't keep up with them, you will start to lose them. It doesn't have to be a lot of work to engage the fans and customers, though. You just need to make sure that the page is getting regular updates, that you are answering questions and comments that your fans place on, and put new videos and pictures up on a daily basis.

Remember with this one that the quality is going to be better than quantity. Posting too many pictures and videos all the time will just annoy some of your followers so keep it down to just the ones that are high quality and try to spread them out to different times during the day to catch fans when they get on. Keeping the pictures and posts to no more than one per a six-hour period will help.

Increasing your presence on your Instagram account is one of the first steps that you will need in order to make more people see about your business. You can always start with some of the fans that you have from other pages, but interacting with some other pages that are similar to yours and making your page look really nice and on topic can help you to really bring in some of those fans that will make your business do better.

Getting More Followers without Having to Spend Money

Now that we have some ideas on how to establish a presence online for your business with the help of Instagram, it is time to work on how to increase your followers. You also need to learn how to expand out some of your networking so that you can find some more followers. Unfortunately, when most beginners start looking at methods of doing this, you will find that the first methods that show up will cost you a lot of money. You can choose to go with one of these "Instagram experts" if you would like to spend a lot of money, but often a new business is on a tight budget and these experts can be overpriced sometimes.

In this chapter, we are going to take a look at some of the techniques that you can use in order to grow some of that fan base without having to spend a whole lot of money. The major advantage of these is that they actually work without having to hire the experts and hope that they work. Some of the techniques that you can try out include:

- *Pick the right time to post. Experts state that posting either at two in the morning or five in the afternoon are the best time periods to post your content and be seen by the highest amount of people.*

- *Use the filters on any of your marketing campaigns for Instagram. Some of the most recent reports show that "Mayfair" is a great filter for a marketer to use on Instagram.*
- *Include a lot of information on the bio of the account. This is not the time or the place to get lazy. You need to make sure to use a lot of information here so that the visitors know what you are all about. Add in some keywords and hashtags as well as the URL of your site. This helps interested users be able to access the website of your business easy if you set it up this way.*
- *Pick out some hashtags that your followers will like. These are always changing so you need to make sure that you check up on this and find the ones that your customers will like.*
- *Figure out the target audience that you would like to use and make sure to target them. This means to find and like their pictures, make comments, and interact them. This is a great approach to use when you want to find some of the best followers.*
- *Hold a contest on Instagram. Most businesses by promoting their event with an image and then they will ask some other users to join in this contest by liking the image. This increases your reach quite a bit if it gets popular and you only have to pay the costs of whatever you are giving away.*
- *Offer some coupons, sneak peeks, and other specials that are offered on Instagram and show them to your Facebook and Twitter followers.*

- *Make sure that you are concentrating on photos that are high quality. Low-quality pictures are going to be a big turn off. You should also consider placing these pictures on Wednesdays because this one is a good day for post engagement.*
- *Pictures that have faces in them are often more popular compared to those that don't have faces so try to do this.*
- *Try to make some relationships happen between you and some of the more influential users on Instagram. This makes it easier for some of the customers that you want to find you and be able to follow you later on. Make sure this is a positive relationship, where you often show up and comment and like some of their posts.*
- *Hold a marketing campaign with some of the other Instagram users. There are often other businesses who are complimentary to your industry and the two of you will be able to work together, with similar promotions helping each other out, to increase your business.*
- *Tag some other people in your pictures. There are two advantages that come from doing this. It is going to first display these pictures in the Instagram feed of the person and it is going to help make it more likely that the photos will be shared.*
- *It is also important to spend some time on the pictures that you are using and sharing on your site. You want them to be high quality, picking out ones that your users are sharing, add*

something inspirational to them, or make some changes so that they tell a story and look nice. Good pictures can help to attract the people that you want while the bad pictures can turn them away.

These are just some of the methods that you can use in order to take your Instagram account and attract some of the viewers that you want. While there are some experts that will charge you a lot of money in order to create your marketing campaign, there are a few things that you are able to do on your own that will help to grow your website and helps you to get the fans and the sales that you want. Keep in mind that this does take a little bit of time, you have to do some of these things consistently over the long term rather than just hoping that it happens overnight, but if you are able to do this, it is easier to get the results that you want.

How to Market the Brand and Increase Sales

While there are many people that will get an Instagram account in order to show some of their own personal things and their day to day life online, you can also use this social media site in order to show off your business and as part of the marketing campaign. There are a lot of things that Instagram will be able to do to help out your business, including to help your brand get out to more people, helping you to get more followers, and helping you to find some good paying customers.

According to some online marketers, Instagram has already surpassed Twitters in terms of how many users are on the social media site each month. This means that you will be able to get your account set up on Instagram and engage with your customers, knowing that they are more likely to come back than on other social media sites. In this chapter, we are going to take some time to look at how you are able to use Instagram for sales and marketing purposes and some of the techniques that you can use to make this happen.

Create a separate account for the business

If you are looking to use Instagram for your business, you need to make sure that it is not mixed with your personal account if you have one. This can confuse your customers if you do try to make your personal and business account into one. You need to make sure that there is an account that is just for your business and that will make the company look great.

In this section, make sure that you are not focusing on yourself. Your personal site can be great for pictures of vacations and all the fun things that you get to do, but there is no place for this on your business Instagram page. It is best to leave these pictures and thoughts for that personal account and focus instead on things that relate to your business on this new page.

In order to make sure that your business account is as effective as possible, here are a few things that you should try out:

- *Be consistent with the profile picture and your name. You need to have relevance and consistency when it comes to using this as a marketing tool and you shouldn't be able to confuse the fans by what is there. This is why using the same name and picture on all of your social media platforms can help to avoid confusion.*
- *Make sure that you have information about your website somewhere on the page. The bio is going to be really great for putting this in because Instagram can be strict about this and the bio is about the only place that this will work on Instagram. While you are at it, make sure that the bio is informative and catches the attention of your fans.*

- *Inside the bio, make sure to list the name of your business at least a few times. Don't let the bio get too long, but make it light and catchy without a lot of sales talk inside of it.*

Share the content that others want to see

For the most part, your users would rather use pictures rather than written words. Even some of the simplest pictures can be effective at spreading your message quickly. And Instagram is really focused on images so it is a good idea to use these. Instagram doesn't usually allow for a lot of long articles and blog type materials because it would rather that you post videos or photos and this is going to work out great when you are working on your marketing campaign.

Many shoppers like to use Instagram because it does focus so much on pictures. This makes it easier for them to take a look at the products that they are interested in and even compare it to some of the other offerings that are on the market. They don't want to read a long blog about the item; they would rather get a chance to see good pictures of the product so make sure that a good deal of your page will focus on these pictures and videos.

Another thing that you should work on is avoiding the hard sell. Most buyers that are on social networking sites are there to collect some information before they make the sale. Many of these are going to be influenced by the activities that are on the social media of the business. If this is good, you will get more sales but if it is bad, you will lose them. Using pushy captions on your pictures is not going to work. Just focus on sharing some great pictures of what you have to offer and spend the captions describing the item. Never force the audience, just give them the information that they need to make the purchase.

There are also a few editing tools that are built into Instagram and can be great for helping you to make your pictures stand out. there are millions of these photos that are added to the platform all of the time and this makes it hard to fight against the competition. You can use some of these filters in order to make the images look better so that they begin to stand out.

Some businesses like to offer some promos to their Instagram followers. There was recently a study done that showed how 41 percent of users on Instagram were willing to follow a business in order to get the promotions and giveaways, even if they hadn't purchased from the business in the past, making it a good incentive to offer these to your followers. Try to make it something fun and make sure that your followers know that this is an exclusive deal just for some of your Instagram followers.

Reach more of your audience

While uploading some good photos is a good way to reach some of your audience and can make it easier to get the results that you want, you also need to do a bit work and have a plan for others to view the photos and follow your page there are many things that you will be able to do to make this happen.

First, hashtags are going t be great because they make it easier for others to see your photos. Most of your fans are gong to see that their feeds are quickly changing so it is easy for the pictures may get buried down before the followers are able to see them when they are away. Using hashtags is going to increase the lifespan of these posts. It is so effective that at least 88 percent of posts on Instagram are going to have at least one hashtag. Of course, you should be careful when picking out the one that you would like to use, picking options that relate to the business and the product, and make sure that you aren't using them excessively.

Look for some brand ambassadors

Forming up a group of ambassadors who are able to share information about your brand with some other users is a great way to reach more people. You can ask some of your followers to share their reviews and their pictures so that you increase your exposures. In some recent surveys, you will find that 78 percent of buyers are going to make their purchases based on the brand and the presence that it has on social media. Thus, when you ask your followers to praise the services and products that you have for sale, it can lead you to some more.

The process for finding these ambassadors can be simple. You will need to find some good hashtags for all of your marketing campaign, encourage other customers and followers to post their own photos and reviews as they relate to your brand. And then when you find out which people are doing this, make sure to reward them in some manner, whether it is through shout outs, discounts, or freebies of some sort.

Encourage interaction with the content

If you want to increase your presence online, you need to make sure that your followers are engaged. Inside of Instagram, you are going to get about 331 engagements if there are around 10,000 followers who are willing to share a photo. This is much higher than some of the other social media sites, but there are some other things that you can do to increase this number including:

Write out some active captions. This can include a call to action, a question, or information about the product if you would like.
Hold a contest: this will help to increase your engagement because people will want to get in on the rewards and they will feel valuable. Make sure to pick out the prize ahead of time and find a way to encourage people to

follow. For example, you can choose to have people like some of the pictures that you have to get entered into the contest.

Respond to your followers: if your followers are asking questions online and you never respond, you will find that it is difficult to keep them around. You need to reply to their comments, answer their questions, follow them, or thank them to make them feel valuable.

Measure how successful the Instagram campaign

As you are working on your campaign, you will want to make sure that you are taking care to track how the campaign is going. If you just throw a few things out online and you don't keep track of the campaign, you will find that things could be going horribly wrong and you have no idea. Placing some measurements in place can help you see how the campaign is doing and make changes if something doesn't seem to be going right.

Make sure that you create a few checkpoints throughout your campaign to help you to check up on things. You can always experiment with something new as well, but make sure that you check out the measurements to see if this is actually successful for you or not. As a new business, you never know what is going to work for you and what isn't going to work so take the time to try out a few things and see what works. Over time, you will be able to see some great results and can stick with the things that work.

Marketing your brand on Instagram can be really successful especially when it is compared to some of the other social media sites, but you need to make sure that you are doing it the proper way so that you bring out the customers and the followers that you would like. Follow some of these steps and you are all set to go with a great marketing campaign.

Getting Your Pictures to Work Well on Instagram

As we have talked about a little bit, it is a good idea to work on taking great pictures to use on Instagram. This is a very visual site and if you aren't good at taking pictures, or you don't have a smartphone that is able to take some decent pictures, you will fid that it is really hard to get the views and the followers that you would like on Instagram. You are competing against many other businesses who want the same kind of customers as you do, so taking care of the pictures that you are posting and making sure that they look nice will make a big difference in the response that you get. Here we are going to take a look at some of the things that you can do to make your pictures look great.

Go with high quality

This does not mean just high quality on how the picture looks, but that it is actually relevant to the information that comes with your business. There are lots of cool memes out there or you may be tempted to show off that latest vacation that you took, but if these have nothing to do with the products and services that you offer, it is going to deflect from the purpose of your page.

With pictures, this should be pretty easy to accomplish, but it is sometimes harder to make a good video that is going to match up. Maybe make a video about a new promotion that you are doing, one about you making one of the meals that your restaurant sells or something similar. You can have a lot of creativity with this one, but if you are talking the whole video about your favorite vacation spots when you sell house insurance, you probably aren't meeting the needs of your customers.

Get the right lighting and angle

First, we are going to take a look at some of the lighting. You have to be careful about some of the lighting that you are using in your pictures. If you take the picture in a dark room, your customers are going to have a hard time figuring out what is going on inside of the picture. Too much light can be hard (so be careful of the flash) because it whitewashes everything that this one is really hard on the eyes. If you are unsure about the lighting, it is a good idea to take a few pictures and see how the item looks when the picture is done.

The angle that you take the picture can really change things up as well. Many pictures are taken with the straight on approach, which can be kind of boring. Taking the picture from above, going from below, moving it around, staging it with other items, or doing something else can add some interest to the picture and may attract the attention a bit more. Consider a few different backgrounds and angles until you are able to find the right one for your pictures.

Do some editing

Remember that you are using Instagram as a business website, so the pictures that you upload will need to look nice and neat. It is not a good idea to just take a quick picture that is blurry or all over the place and assume this is going to bring in the customers that you are looking for.

There are a number of edits that you can do to make the pictures look better. First, get rid of any of them that are blurry and just don't look that good. You may find that cropping the picture to get rid of some of the background noise can help it look better, changing some of the light tones, and doing a few other quick edits can help to make the pictures look more professional.

Use the tools on Instagram

The nice thing about Instagram is that it offers some of its own editing tools that you are able to use. This means that once you upload the picture, you will be able to mess around with it even more in order to make it look great. You can add some borders, change it around so that it fits into the little area that you are given, change up the colors, or do some other cool things on the pictures.

Sometimes you will want to just keep the picture the way that you created it, simple and just showcasing your business, but other times it is a great idea to change some things up and make them look unique. It is worth your time to spend a bit of time looking through the tools that Instagram has to see how they can make changes to your pictures.

Post at the right time

In addition to adding some little things to the picture to make it look nice, you also need to worry about the time that you will post these pictures so the highest amount of people will find them. This is sometimes going to be an activity of trial and error. As you post pictures on your sites and take a look at the analytics that are taken, you will start to notice that there are some trends for when your customers are most likely to be online and engaging. Once you learn what these times are, you can post more and really take advantage of this.

If you are just getting started with Instagram, it can be hard to figure out what times are the best for posting. There are some Instagram professionals who have found that posting at 2 in the morning or 5 in the afternoon can be the best times because these often show the highest engagement numbers and could help you to reach the customers that you want.

Now, if you find that these times don't work for you (how many business owners are ready to post at 2 in the morning) it is fine to post at different times and find the right time periods for your business. And it is even a good idea to post at different times during the day. Just don't get too carried away with this. Your customers don't want to see hundreds of pictures being posted each day because this gets annoying. You may post a number of pictures when you are first starting in order to fill up the portfolio and your page, but after that, a few posts a day is about all you need, especially with the right hashtags. Most experts recommend going at least six hours in between your postings to prevent any annoyance with your customers.

Instagram is a great option to use if you would like to start meeting with more customers and growing your business, but you must remember that this is a very visual social media site and if you don't have high-quality pictures and videos, you will end up turning away some of the customers that you are working towards. Make sure to follow some of these great tips and you will put up meaningful posts that your customers are looking for.

How to Successfully Interact with Others on Instagram

Interacting with others on Instagram is a great way to get some more of the customers that you want. You will be able to interact with the followers that you get as well as interact with some of the other businesses and personal pages that are out there. You need to have a plan in place for this interaction, but reaching out to others and sharing information and more together will help both of you to gain some of the customers that you need. Here are some of the simple steps that you can take in order to increase your followers with the help of Instagram interaction.

Invite your other followers to Instagram

If you have a few other social media sites, you can talk to some of your followers from there and invite them over to your Instagram page. Some of your customers are going to enjoy this option better because that is the one that they use and others will do it in order to get some extra deals and such compared to the other sites.

Starting with your current followers on other social media is a good place to begin, but you may not get all of them and even if you do, they were already your customers and it isn't really adding in any new ones to the mix. Adding information about your Instagram account on your other marketing materials so that new followers can find out information and learn about the products and services that you have to offer.

Respond when someone comments on your posts

As you get more followers, there are going to be times when they will comment on your posts. This is a good thing because it can help to bring your information to the forefront of searches and shows that your customers are interested in what you are posting. But it is your job to post back to what they are saying. If they have a question, answer back with the answer that they need. If they leave a suggestion or just a simple comment, make sure to respond to that as well.

Ignoring the things that your customers post on your page looks unprofessional. It shows that you may not be on your site as much as you should be or that you don't value your customers that much because you can't take more time than posting to talk to them. You don't have to be on the account all the time to do this and you don't need to respond right away to each one. Pick a time once or twice a day when you will get on and respond to as may of the customer and follower comments on your page.

Have promotions and sales

There is nothing that your followers like more than a good deal and adding some promotions or sales can be a great way to get them to interact with you and want to check out your site more. The first thing that you can try includes sales that are just for your Instagram followers. List up the sale both on Instagram and on other social sites, as an exclusive to those who are following you on Instagram. This will show appreciation to those who are already following you on Instagram and it could also bring some more people over to your page as well.

There are many types of promotions that you can choose to run with, but one that works well is a contest or a drawing. Many Instagram businesses will offer some kind of discount or prize to those who like and share some of their posts. This can include a drawing to decide who gets it or you pick the 100th person who does this with the picture. This brings out some excitement on the page to win the item and it helps for more people to see your information as the picture gets shared and liked many times over.

Add some useful comments to other pages

Interacting with other pages is another great way to bring in more customers. Find a few pages that you really enjoy as well as a few that relate in some way to your business (they don't have to be exactly the same, but something similar so you have some of the same customers) and start to follow them. But don't just be a passive follower on their pages. Like the pictures and videos that they post and even leave some comments as well.

Now, you need to take some time to actually write out good comments on these pages. It is not enough to write something like "nice job!" on a post and get noticed. These are the most basic types of comments and are usually ignored by the poster and their customers. You need to add some value to the page, writing out at least a few sentences to explain what you like or how you think the other person can improve. These more in-depth comments are going to get more interest and can lead more customers back to your page.

There are a few things to remember here. First, don't add in information about your own business in the comment. People can see the name of your company in the posting already and if they are interested in checking you out, they will. Adding in a comment about your sales is a bad tactic and can turn some people off. Second, make sure that the comment adds something of value to the poster and their audience so really think out the things that you are posting. And finally, this is not a place to make fun of the poster or bring them down. Keep the comments positive and happy with some good information for the audience, without digging into the poster, and you will see some better results.

Try not to add too much spam or sales to interactions

As a beginner, you may be really excited to get your business started and to see some success with your marketing plans. But you need to remember that you can't be too salesy when talking to other people. If you just go over to another page and start writing that others should come and visit your site. When you comment, you need to add some value to the other page and to their customers if you want to bring them over, or you will just be seen as a spammer. The good news is that while these comments may take a little bit of time to create, customers will actually notice the work and will check out your store.

Interacting with others on Instagram is one of the best ways that you are able to bring in more customers. You can't just post a few pictures and videos on your account and assume that people will want to learn more about you and visit your store. You need to get out there, let others know that you are there and what you have to offer, and interacting with your customers and other pages on Instagram will help you to get this done. Make sure to use the tips above will make it easier than ever to see the great results that you want.

Common Mistakes to Avoid When Marketing on Instagram

Many beginners are excited to bring their business onto Instagram and share some of their products and services with others online. Instagram is great to use and it is good that you are excited, but you need to make sure that you are doing it all in the right way. There are some mistakes that you could make that would drive away the potential customers and could make all of that hard work that you are doing go to waste. Some of the common mistakes that you should avoid with Instagram include:

Combining your business and personal profiles

Sometimes it is tempting to combine the personal and business profiles on Instagram together, but this is going to take away from the professionalism of your page and can sometimes confuse other people about what the page is all about. If you want to have a personal page that shows your daily life and all the cool things that you do and see, then set this up separate from the business page that you are trying to create.

Your business page should only have stuff that has to do with your business on it. This helps your customers to know what to expect and what kinds of products and services that you offer. There is nothing wrong with using a personal account for some of your personal life stories, but this should never combine with your professional or business page.

Ignoring the idea of sponsored posts

If you would like to be able to gain a lot of exposure for your business page, going with sponsored posts can be a great way to do this. In order to leverage your sponsored post and get all the benefits, you must find some accounts on Instagram that are more influential and which have quite a few followers already in place; you also want to make sure that these sites that you are going with will have the same target demographic and niche that you want to work with as well. Then you need to get in touch with some of these users and find a way to convince them that it is a good idea to promote some of your content on their news feed.

Now, this can be a hard task to get done. What do you need to do in order to get these people who have influential accounts to take the sponsored posts, especially if they already have lots of people asking the same thing? For the most part, you should check the bio of the accounts that you want to work on because these are usually going to mention whether they accept these sponsored posts or not and they will include their contact details to help make it easier to communicate with the parties that are interested. If you see something like "Open to Business Enquiries" or "Accepts Sponsored Posts Requests" in the bio, make sure to email your details to the account that you are interested in and ask to see what their pricing structure is; you are also able to do some negotiations if you would like.

It is going to cost you a little bit of money to use this technique, but it is one that has been proven to add thousands of new followers to your account in just a short amount of time. There are many marketers on Instagram that have used this strategy to reach out to a lot of customers who are not already their followers. As a beginner, you may feel that you shouldn't spend the money on this, but considering the amount of followers it can add to your Instagram page, it is well worth your time and money.

Improper use of hashtags

Hashtags can be an important tool that you can use on Instagram, but you need to make sure that you are using them and using them properly. If you are sending out Instagram posts without a hashtag, you are really missing out on an opportunity to get some new users.

On Instagram, you are allowed to add up to 30 of these hashtags so go ahead and use a few. You don't want to go overboard and you probably don't need to use all 30 because it does look a bit spammy and can make you lose some of your credibility. But there is nothing wrong with adding in two or three of these with each post is going to help you to see some results and bring in more of your customers.

Forgetting to respond to user comments

It is easy to get caught up in all the things that you have to do to keep your business running, but if you want to interact with your customers and make them feel valuable, you need to respond to the comments that others leave on your posts. Even if it is something as simple as thanking them for responding or you answer some of their questions, this shows the customer that they are important. Also, make sure to be there not only before the sale but afterward as well. Some customers have questions after they make a purchase and no one wants to feel like they are just a number because you ignore their questions or their concerns.

Checking your account a few times a day to look at the comments and respond as needed. You don't need to be on all the time and customers don't expect you to be there instantly all the time, but having a regular schedule for checking in on the account and assisting your customers can help you see better results.

An incomplete bio

Inside of Instagram, you will get a small space to write out some details about your business and the brand that you are working with. This is an important part of your business because it is going to help to show you as a professional in the industry and gives potential followers a little look into what you provide.

As a business, you need to fill out your whole profile, including the bio. People are not going to follow you if the profile is not fully complete because it shows that you didn't really value this information or helping out the customer. Another mistake that beginners can make? Forgetting to place their website link inside of the bio. You will not be able to place this link with your pictures

The Best Tips for Growing Your Instagram Business Account

Now that we have had some time to learn what Instagram is all about and all of the great ways that it is able to help us to add to our marketing plan without having to spend a lot of extra money on a social media expert. But even though we need some of the basics, what are some of the things that you are able to do in order to make yourself stand out on Instagram so that your potential customers would choose you over someone else. Some of the best tips that you can do in order to help grow your Instagram business account include:

- *Connect Facebook and Instagram together: these two social media sites are actually connected and when you let them work together, you will find that they can be really powerful tools. Make sure that Instagram and Facebook are connected so that you can get the best from both of them and boost some of the marketing efforts. You can even add in a tab for Instagram on your Facebook page so that when you post something on Instagram, it will automatically link up with the Facebook and send it over, saving you time.*
- *Create your strategy: as a business owner, you need to make sure that you have a good brand strategy that is going to help your business to grow. You will want to make sure that whatever strategy you start to use on Instagram is going to keep the focus*

on the brand that you have built and how that brand sees the world. Instagram is great for sharing videos and photos so make sure that when you connect the business with your followers, make sure that it stays consistent with your brand rather than straying away or showing things that just don't go together.

- *Use your brand or company name in hashtags: the hashtags that you pick don't have to be complicated. If you already have a pretty good following on Instagram, or your brand name is well known, go ahead and use this as one of your hashtags. This will make it easier for your followers to find you because they can just search the name and find some of your posts.*

- *Make a follower famous: it is not just about your followers checking out your posts, it is about how you interact with your followers. Take the time to look over the pages of your followers and then like and share some of their posts. This helps to show the customers that you really appreciate them because you are acknowledging their cool posts and sharing them with others on your page. Be careful with this one though because some may be personal and you should always ask for permission beforehand if you are unsure whether they would like it shared or not.*

- *Be creative with the pictures: it is not enough to just take a few pictures of your products or services and then call it good. You need to get a bit creative. There are thousands of pictures of food or sweaters and so on, how are you going to make your pictures draw the attention of the customers that you want? You can use filters, different angles, changing the lighting, and some of the other tricks on photography to help you out. you don't need to hire a professional photography to do this, but you should be a bit creative with what you are doing.*

- *Work on videos: a new feature that you are able to find with Instagram is that now you can add videos to your postings. This is a new way to interact with your customers, telling them about some of the products, bringing up rules about a promotion, or doing something else that is creative. Just like with the photos that you post, the videos need to have something to do with the business, but it is fine to be a bit creative with this.*
- *Think about how to show your products: remember that as a business, you need to make sure that you are showcasing some of the products that you sell. Show these products in a bunch of different ways. This is a mobile site so be creative when showing it off, but many times businesses see success when they show the products in real life, such as something wearing the sweater on a walk rather than just having it lay there.*
- *Post some good videos (or you): if you are the CEO of your business, this means you need to post a few videos of yourself. This is meant to make the top executive of the company look personable. Make some short and even some quirky videos of the*

CEO to add to Instagram. Showing some of your hobbies outside of work, for example, can be a lot of fun.

- *Partner with some other brands: this can be beneficial for both you and for the other brands that you decide to work with on this promotion. If the two of you are reaching for similar customers but are different products, it can make it easier to find customers that both of you can use. Partner up with a few other brands and have them post some of your products on their feeds while you do the same for them.*
- *Post on a consistent basis: it is not going to do well for your business if you just post for a few days here and there, ignore the profile for awhile, come back for a week, and then ignore it again, keeping up on this same cycle for the next few years. If you want to find more followers and get more customers, you need to post with some consistency. You can even post some of the same things on Instagram more than once, just make sure that you find a good rhythm to when you post. Sure there will be days when you miss out because you get busy, but make sure to keep up with this as much as possible.*
- *Track your results: how are you supposed to know if what you are doing is successful if you aren't keeping track of what is going on with analytics and other tracking tools. You should use some of the tracking tools that are available through Instagram and monitor when is the best time for you to post, what your followers*

like the most, and other things that make it easier for you to get better results in the future.

- *Host a photo contest: there are many options that you can stick with when it comes to having a photo contents. The first option is to have your followers post some pictures of them using your products, and then have a vote on which ones are the best with the winner getting a prize. Sometimes you can post one of your own pictures and have your followers like and share that picture to be entered for a prize as well. Videos can work in much the same way. You can mess around with this one a bit to see what you are able to figure out for terms and prizes, but this is a great way to get people to interact a bit more on your page and to share your products even more.*
- *Keep up with the trends; the trends on social media sites will often change over time. Things that work today may not work as much later in the future. If you refuse to keep up with some of these trends, you are going to start missing out on some of the customers that you need. In some cases, missing out on trends could end up hurting you in trends and search results. The trends can be beneficial to you when it comes to helping your customers and getting ranked higher in the site's search.*

Working on Instagram does not have to be something that is difficult or impossible to do, as long as you make a plan and try to showcase your work while interacting with your customer. It will take a bit of time to make this work, but overall, you are going to be able to put your Instagram account to work for you and it will bring in a great conversion rate so your products begin to sell.

Conclusion

Thank for making it through to the end of Instagram Marketing: Social Media Marketing with Instagram, Gain More Followers and Monetize Your Instagram Account. Let's hope it was informative and able to provide you with all of the tools you need to achieve your goals of

The next step is to get online and start your own business account on Instagram. If you already have a personal account, it is best to start one that is just for business, or completely restart the personal one so that it is ready for all your business needs without mixing it with personal. Once that is done, it is time to start posting high-quality pictures to show off your products and your services, reaching out to potential customers, and bringing out your presence on this social media site.

This guidebook spent some time looking at the benefits of using Instagram marketing and why you would want to choose this for growing your business. We talked about some of the benefits of getting started with Instagram, how to start your own account, the importance of having good pictures, and some of the ways that you can attract some more followers to your page. In the end, you will be able to use Instagram to really bring in the customers without having to pay a lot of money to an Instagram expert.

When you want to expand your social media marketing campaign and you want to really get the high sales and engagement rates that come with Instagram, make sure to check out this guidebook and learn as much as you can about how to market your business on Instagram.

Finally, if you found this book useful in any way, a review on Amazon is always appreciated!

© Copyright 2017 by Elizabeth Nightingale *- All rights reserved.*

The following eBook is reproduced below with the goal of providing information that is as accurate and as reliable as possible. Regardless, purchasing this eBook can be seen as consent to the fact that both the publisher and the author of this book are in no way experts on the topics discussed within, and that any recommendations or suggestions made herein are for entertainment purposes only. Professionals should be consulted as needed before undertaking any of the action endorsed herein.

This declaration is deemed fair and valid by both the American Bar Association and the Committee of Publishers Association and is legally binding throughout the United States.

Furthermore, the transmission, duplication or reproduction of any of the following work, including precise information, will be considered an illegal act, irrespective whether it is done electronically or in print. The legality extends to creating a secondary or tertiary copy of the work or a recorded copy and is only allowed with express written consent of the Publisher. All additional rights are reserved.

The information in the following pages is broadly considered to be a truthful and accurate account of facts, and as such any inattention, use or misuse of the information in question by the reader will render any resulting actions solely under their purview. There are no scenarios in which the publisher or the original author of this work can be in any fashion deemed liable for any hardship or damages that may befall them after undertaking information described herein.

Additionally, the information found on the following pages is intended for informational purposes only and should thus be considered, universal. As befitting its nature, the information presented is without assurance regarding its continued validity or interim quality. Trademarks that mentioned are done without written consent and can in no way be considered an endorsement from the trademark holder.